P9-CSU-615

¡Criaturas diminutas!/Bugs, Bugs, Bugs!

Grillos/Crickets

por/by Margaret Hall

Traducción/Translation: Dr. Martín Luis Guzmán Ferrer
Editor Consultor/Consulting Editor: Dra. Gail Saunders-Smith

Consultor/Consultant: Gary A. Dunn, MS, Director of Education
Young Entomologists' Society Inc.
Lansing, Michigan

Mankato, Minnesota

Pebble Plus is published by Capstone Press,
151 Good Counsel Drive, P.O. Box 669, Mankato, Minnesota 56002.
www.capstonepress.com

Printed in the United States of America

1 2 3 4 5 6 11 10 09 08 07 06

Library of Congress Cataloging-in-Publication Data
Hall, Margaret, 1947–
[Crickets. Spanish & English]
Grillos = Crickets/de/by Margaret Hall.
p. cm.—(Pebble plus. ¡Criaturas diminutas!/Bugs, bugs, bugs!)
Includes index.
ISBN-13: 978-0-7368-6678-1 (hardcover)
ISBN-10: 0-7368-6678-7 (hardcover)
1. Crickets—Juvenile literature. I. Title. II. Title: Crickets. III. Series: Pebble plus. ¡Criaturas diminutas! (Spanish & English)
QL508.G8H35 2007
595.7'26—dc22 2005037467

Summary: Simple text and photographs describe the physical characteristics and habits of crickets—in both English and Spanish.

Editorial Credits
Sarah L. Schuette, editor; Katy Kudela, bilingual editor; Eida del Risco, Spanish copy editor; Linda Clavel, set designer; Kelly Garvin, photo researcher; Karen Hieb, product planning editor

Photo Credits
Bill Johnson, 20–21
Bruce Coleman Inc./Carol Hughes, 17; David C. Rentz, 18–19; Janis Burger, 11; John Shaw, 15; Raymond Tercafs, 4–5
David Liebman, 1, 12–13
Dwight R. Kuhn, cover
James P. Rowan, 8–9
Robert McCaw, 7

Note to Parents and Teachers

The ¡Criaturas diminutas!/Bugs, Bugs, Bugs! set supports national science standards related to the diversity of life and heredity. This book describes crickets in both English and Spanish. The images support early readers in understanding the text. The repetition of words and phrases helps early readers learn new words. This book also introduces early readers to subject-specific vocabulary words, which are defined in the Glossary section. Early readers may need assistance to read some words and to use the Table of Contents, Glossary, Internet Sites, and Index sections of the book.

Table of Contents

Tabla de contenidos

Crickets

What are crickets?

Crickets are insects that leap.

Los grillos

¿Qué son los grillos?

Los grillos son insectos

que brincan.

How Crickets Look

Most crickets have black, green, or brown bodies.

Cómo son los grillos

La mayoría de los grillos tienen los cuerpos negros, verdes o marrones.

Crickets are about the size of a lima bean.

Los grillos son como del tamaño de una haba.

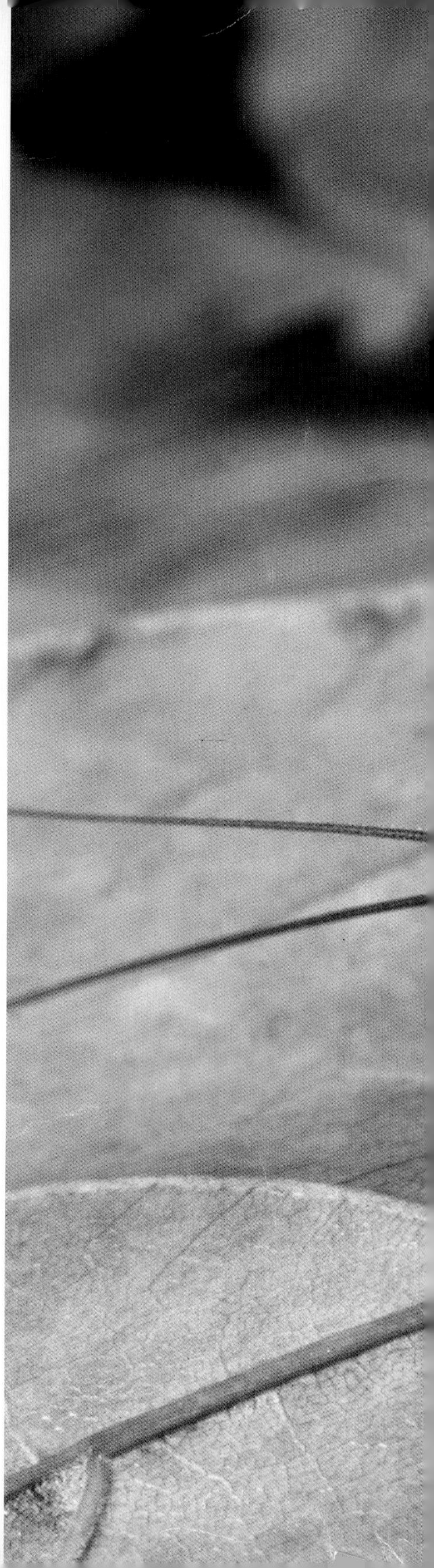

Crickets have six long legs.
They jump and hop.

Los grillos tienen seis patas.
Los grillos saltan y dan brinquitos.

Crickets have sharp jaws.
Jaws help crickets bite
and chew.

Los grillos tienen mandíbulas
filosas. Las mandíbulas los
ayudan a morder y masticar.

Crickets have two antennas. Antennas help crickets feel and smell.

Los grillos tienen dos antenas. Las antenas los ayudan a sentir y a oler.

What Crickets Do

Crickets hide during
the day. They come out
at night to eat.

Qué hacen los grillos

Algunos grillos se esconden
durante el día. Salen por
la noche a comer.

Male crickets chirp. They rub their wings together to make chirping sounds.

Los grillos macho cantan. Para cantar frotan sus alas una contra otra.

Crickets listen to each other. They have ears on their legs.

Los grillos se escuchan los unos a los otros. Tienen oídos en las patas.

Glossary

antenna—a feeler; crickets use antennas to sense movement and to smell.

insect—a small animal with a hard outer shell, six legs, three body sections, and two antennas; most insects have wings.

jaw—a part of the mouth used to grab, bite, and chew

male—an animal that can father young

Glosario

la antena—parte del cuerpo para tocar y sentir; los grillos usan las antenas para sentir el movimiento y olfatear.

el insecto—animal pequeño con un caparazón duro, seis patas, cuerpo dividido en tres secciones y dos antenas; la mayoría de los insectos tiene alas.

el macho—que puede ser padre

la mandíbula—parte de la boca que se usa para atrapar, morder y masticar

Internet Sites

FactHound offers a safe, fun way to find Internet sites related to this book. All of the sites on FactHound have been researched by our staff.

Here's how:

1. Visit *www.facthound.com*
2. Choose your grade level.
3. Type in this book ID **0736866787** for age-appropriate sites. You may also browse subjects by clicking on letters, or by clicking on pictures and words.
4. Click on the **Fetch It** button.

FactHound will fetch the best sites for you!

Sitios de Internet

FactHound proporciona una manera divertida y segura de encontrar sitios de Internet relacionados con este libro. Nuestro personal ha investigado todos los sitios de FactHound. Es posible que los sitios no estén en español.

Se hace así:

1. Visita *www.facthound.com*
2. Elige tu grado escolar.
3. Introduce este código especial **0736866787** para ver sitios apropiados según tu edad, o usa una palabra relacionada con este libro para hacer una búsqueda general.
4. Haz clic en el botón **Fetch It**.

¡FactHound buscará los mejores sitios para ti!

Index

Índice